JN241915

Hiro and the Miso Ghost

All about Hatchō Miso

Sankeisha

There's something squiggling in my miso soup.

"Hiro, why don't you drink up your miso soup?"

Mom is always angry with me because I don't want any miso soup.

"I don't want to drink it."

Something is squiggling around in my bowl. It's moving back and forth.

It looks like a face.

Hey, get out of my soup, squiggling miso ghost!

おみそしるの中には、小さなオバケがすんでいる。「ひろくん、どうしておみそしる、のまないの！」ボクがおみそしるのまないから、ママはいつもボクをおこるんだ。「のみたくないもん！」ほら、いまだって、さわってないのに、おわんの中で、なにかが、モヤモヤうごいてる。あ、ボクのことにらんでるカオになった！

"I don't like miso soup. Can't I just have a sandwich instead? On kindergarten days, I always get a sandwich for breakfast. Why can't we have sandwiches for breakfast on the weekend? Why do we have to have rice and miso soup on the weekend?"

"Why not have nothing at all? You think I want to give such a spoiled little boy any breakfast?"

"I don't like Mom any more."

So I run out of the house.

「おみそしるなんて、やだ！　いつものパンがいい」ようちえんにいくときはパンなのに、どうしてお休み(やす)の日(ひ)は　ごはんとおみそしるなの？
「そんなワガママいう子(こ)は、ごはんなしです！」だってオバケなんて、のめないもん。「ママのバカー！」ボクはいえをとびだした。

I know. I can visit Grandma. She lives right next door.

“Grandma!”

“Oh dear, what’s the matter with you?”

“I just don’t like miso soup. So Mom got angry. Again.”

Grandma doesn’t know what to say. Then she asks,

“Why Hiro, don’t you like miso soup?”

Mom might laugh at me, but I can tell Grandma anything.

となりのおばあちゃんちに、いくからいいもん！「おばあちゃーん！」「おやおや、どうしたの？」「おみそしるをのまないから、またママにおこられた」おばあちゃんは、ちょっとこまったかおをした。「ひろくんは　どうしてこんなに　おみそしるがきらいなのかねぇ」ママにいったら、わらわれるかもしれないけど、おばあちゃんなら、おしえてもいいかな。

“Can I tell you a secret? Just between you and me?”
“Ok. Just between you and me. What’s the secret?”
“There’s something squiggling in the miso soup.”
“Squiggling?”
“Grandma, something is squiggling around
in that black soup. It’s alive. I bet it’s a ghost.
A miso ghost!”
“Maybe. But I think I might just know where
the squiggling ghost comes from.”
Grandma thinks a bit. Then her eyes sparkle
a little.
“Come on Hiro, time for a little trip.”

「おばあちゃん、ママにはひみつだよ」「なあに?」「おみそしるの中にはオバケがいるんだよ」「オバケ?」「だって、まっくろなしるの中で、なにかがいきてるみたいに、モヤモヤうごいているんだよ。あれは小さなオバケがうごかしてるに　きまってるよ」「そっか。そっか。オバケのせいなんだ」
おばあちゃんは、すこしかんがえてから、にこっとわらった。「ひろくん、おでかけしましょ」

Off we go. Grandma and I always take a walk on this street.

“Are we going to Okazaki Park?”

“We are going to visit my friend’s place on the other side of Okazaki Park.”

Who could Grandma’s friend be?

とっとこ、とっとこ。いつも　おばあちゃんと　さんぽするみち。「おかざきこうえんに、いくの？」「おかざきこうえんの　むこうにある、おばあちゃんの　おともだちのところよ」おばあちゃんの　おともだちって、どんな人（ひと）だろ。

"Hatchi" is the Japanese word for "eight". Hatchō miso got its name from "hatchi chō". A "chō' was an old measure of a standard city block equivalent to 109.1 meters (358 feet). Maruya Miso is eight "chō" (blocks) from Okazaki Castle, about 870 meters (a little over half a mile).

“Grandma’s friend’s place has a lot of big barrels and mountains of stone. So where in the world are we going?”

「おばあちゃんの　おともだちのいるところは、大(おお)きなオケと石(いし)の山(やま)があるところ。どのみちを　とおっていこうか？」

“Where in the world are we?”

“Over here, Hiro. Welcome to Maruya Hatchō Miso.”

“Hatchō miso?”

“This is the factory where we make hatchō miso.”

“Smells like it, all right. I can smell the miso all right.”

「ここはどこ?」「いらっしゃい、ひろくん。まるやはっちょうみそへ、ようこそ」「はっちょうみそ?」
「ここは、はっちょうみそをつくる　こうじょうだよ」「ほんとだ。おみその　においがする」

"Hiro, do you still think there's a squiggling ghost in the miso soup?"

I get really embarrassed and give Grandma a dirty look. But she just smiles back at me.

"I told you this was our secret. You're not supposed to tell anyone."

"But there's nothing to be ashamed about, Hiro. Miso is alive."

"Huh? Miso really is alive? There really is a squiggling ghost in the miso?

"Let me tell you all about the secret of miso."

「ひろくんは、みそしるには　オバケがいると　おもってるの？」ボクははずかしくて、ニコニコわらっているおばあちゃんを　にらみつけた。「おばあちゃん、ひみつだっていったのに！　しゃべらないでよ！」「いやいや、ひろくんはまちがってない。だって、みそは生きてるんだからね」おみそが生きてる？　やっぱり　オバケがはいってるの？　「おじさんがね、みそのひみつ　おしえてあげるよ」

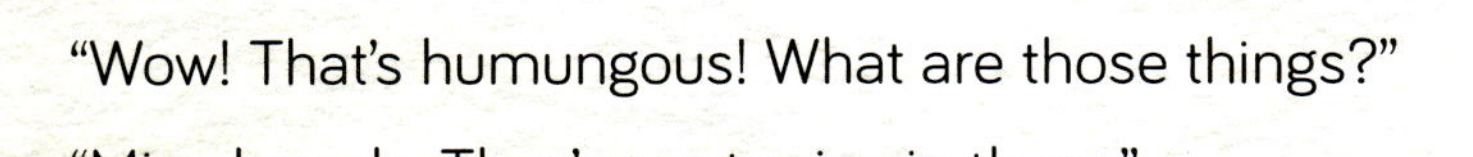

“Wow! That’s humungous! What are those things?”

“Miso barrels. They’ve got miso in them.”

“There’s miso in all these barrels?”

“Yes, but just baby miso at the moment. Now they are growing up into adult miso.”

“So the ghosts are growing up.”

I think that’s kind of cool.

「わぁーーー大きい！　なにこれ！」「これがみそオケ。この中にみそが入っているんだ」「ぜんぶのオケに、おみそが入ってるの？」「そうだよ。このオケの中のみそは　まだ子どものみそ。いま、いっしょうけんめい、おいしい　おとなのみそになるために、がんばってるんだ」「ふーん、オバケががんばってるんだ」ボクはちょっとだけ　すごいとおもった。

"Hatchō miso starts out as just these soybeans and salt."

"These soybeans are like the beans we throw out the window to get rid of bad spirits in our house."

"But they're really hard. These hard beans aren't much like miso though."

"That's right. We use something we call *kōji* to turn these soybeans into miso. *Kōji* is a kind of little friend that gets things started."

"*Kōji*?"

"So *kōji* is that ghost I saw."

「はっちょうみそはね、はじめはこの大豆としおだったんだ」「大豆って、マメまきのときのマメみたい。こんなかたいマメ、おみそとはぜんぜんちがうよ」
「そうだね。このぜんぜんちがう大豆を　こうじがみそにしてくれるんだ」「こうじ？」「オバケの　しょうたいは、こうじっていうんだよ」

"*Kōji* are small invisible living things. When we steam soybeans to make them soft, and give them to the *kōji*, they eat the soybeans and we get more and more *kōji*.

When I think too much about it, it's kind of scary.

「こうじは目に見えないくらい小さい小さい生きもので、やわらかく むした大豆をあげると、大豆をたべて、ドンドンドンドンふえていく」

“We pack the soybeans that the *kōji* are eating into these great big barrels.”

“After that, we pile all these great big heavy stones on top of them.”

“Oh no! What about the *kōji*? Don’t they get crushed?”

“Not to worry. *Kōji* are so strong they can lift these great big stones right up.”

「こうじがたべつづけている大豆を　このオケいっぱいにつめる」
「それから上に石をつむ」「こうじくん、おもくてつぶれちゃうよ」「だいじょうぶ、こうじはこんな大きな石をもち上げるくらい、力がつよいんだから」

It's nice and warm. *Kōji* can eat a lot.
ぽかぽかいいきもち。モグモグモグモグいっぱいたべる。

It's hot, but very nice. There are lots of *kōji* that grow and lift up the stones.
あつくてもゲンキ！ いっぱいふえて石(いし)だってもちあげちゃう！

It's cool and time to rest.
すずしくなったね。
モグモグはすこしやすむよ。

It's cold and *kōji* can hibernate.
ブルブルさむいなー。
ちょっとじっとしていよう。

"Then we just wait around while the *kōji* make our hatchō miso."

"So, Uncle Maruya, the *kōji* do all the work and guys like you just sit around doing nothing? That's not fair!"

Grandma and Uncle Maruya laugh a little.

"You're quite right. I don't do anything I just make sure that the *kōji* make miso. It's something I've been doing for a very long time."

"How long?"

"Since your grandma's grandma's grandma's days."

"I don't get how far back that was, but sounds like a long time ago."

「こうやって　まってると、はっちょうみそができる」「こうじくんは　がんばってるけど、みそのおじちゃん　なんにもしてないじゃん」おばあちゃんとみそのおじちゃんがわらった。「そうだね、なんにもしてない。おじちゃんがやってるのは　こうじがみそをつくりやすいように手(て)つだってるだけ。ずーっとずーっとむかしからね」「どれくらい　むかし?」「おばあちゃんのおばあちゃんの、もっともっとまえのおばあちゃんの　生(う)まれるまえね」「おばあちゃんの　おばあちゃんの　おばあちゃんなんて、もうわかんないよ」

Spring

When it's warm again, *kōji* can eat a lot.
あったかいから　モグモグモグモグまたたべよう。

Summer

We are feeling good again. Chomp-chomp, squiggle-squiggle, yummy-yummy!
またまたゲンキ！　モグモグポコポコ　モグモグポコポコ…

Autumn

When it starts getting cool, we stop eating.
すずしくなって、モグモグするのも　もうおしまい。

Winter

When two summers and two winters have past, we can become delicious miso.
ふたつのなつと　ふたつのふゆがすぎ、ぼくらはおいしいおみそになりました。

"Hiro, I wonder if you'd like to try the miso the *kōji* made."

OK, I get it. It's *kōji*, not a ghost. But it's still kind of scary to me.

"Am I going to get a stomachache?"

"No, you won't. You know, soybeans that get turned into miso are like meat from plants."

"Meat from plants?"

"Soybeans have a lot of nutrition and this makes your body get nice and strong."

"Awesome!"

"The miso made from soybeans is the same, so miso is really good for you."

"Does miso soup work the same way?"

"Yes, of course."

「ひろくん。こうじが　いっしょうけんめいにつくった　みそ、たべてくれるかな」オバケじゃなくて、こうじだってわかったけど、まだすこし…こわい。「おなか　いたくならない?」「いたくならないし、みそをつくる大豆は、はたけのにくっていわれるんだよ」「はたけのにく?」「大豆(だいず)はえいようがあって ひろくんの体(からだ)をつくるもとになる。たべると　ひろくんが大(おお)きくなれるんだよ」「すごい!」「大豆(だいず)からできたみそも同(おな)じ。それに、こうじがつくってくれた　からだにいいものも、いっぱい　入(はい)っているんだよ」「おみそしるにも入(はい)ってる?」「もちろん入(はい)ってるよ」

大きくげんきになる
Big and healthy!

"Here you are. This is konnyaku jelly. We made it from konnyaku yams with miso sauce on top that we made from hatchō miso.
I take a bite of the konnyaku that Uncle Maruya gave me.
"Grandma, this is really good!"
"Yes, I know. Delicious."

「はい、これどうぞ。うちのはっちょうみそでつくった、みそだれつきのこんにゃくだよ」 おじさんがくれたこんにゃくは、すぐにパクパクたべちゃった。
「おばあちゃん　おいしー！」「うんうん、おいしいね」

“We can use hatchō miso in all kinds of dishes like this.”

“Mmm, looks really good!”

「はっちょうみそは、ほかにも　こんなりょうりに　つかえるんだよ」「わーおいしそー！」

“Grandma, I would like to buy some hatchō miso for Mom.”

“OK, we can go down to the store and buy the miso.”

“I want to be healthy and eat lots of *kōji*.”

“Good idea.”

「おばあちゃん、ボク　はっちょうみそ、ママにおみやげにしたい」「じゃあ、おみせでおみそを　かっていきましょうね」「こうじくんをたべて、げんきになりたいな！」「そうね」

"I wonder if Mom can make some kind of yummy food with this miso."

"Of course, she can. But before that, you've got to start drinking your miso soup. It's got *kōji* in it too, you know."

I feel kind of embarrassed, but I hold onto Grandma's hand nice and hard. She's right, you know.

「ママ、このおみそで　おいしいりょうり、つくってくれるかな？」「きっとつくってくれるわよ。でもそのまえに、こうじくんのはいった　おみそしるをのまなくちゃね」ちょっとはずかしかったけど　ボクは、おばあちゃんの手を　ギュッとにぎりしめた。

Maruya Hatchō Miso

52 Okan-dori, Hatchō-chō,
Okazaki City, Aichi Prefecture, Japan
Tel: 0564-22-0222
(from outside Japan +81-564-22-0222)
www.8miso.co.jp/english.html

Want a tour of the factory?

Entrance: 9:00am - 4:20pm
There are guided tours on the hour and at 30 minutes past the hour from 9am to 4pm.
Visitors' reception
Telephone: 0564-22-0678 (from outside Japan call +81-564-22-0678)
E-mail: info@8miso.co.jp

How to find us:

1 minute on foot from Okazaki-kouen-mae Meitetsu train station.
10 minutes by taxi from Higashi Okazaki Meitetsu train station.
15 minutes by car from Toyota Higashi IC expressway junction (Tōkai-Kanjō & Isewangan Expressways).
15 minutes by car from Okazaki IC expressway junction (Tōmei Expressway).

Hiro and the Miso Ghost
All about Hatchō Miso

2018年 5月1日　初版発行
2018年 7月1日　2刷発行

定価（本体価格1,800＋税）

作：すずき あきこ　絵：ひび たかあき
監修：金城学院大学　生活環境学部　食環境栄養学科
教授：丸山 智美
製作協力：安藤 竜二
英語版：クレイグ・アラン・フォルカー
（ジェームズ・クック大学，オーストラリア）
寶壺 貴之
（岐阜聖徳学園大学 経済情報学部 准教授）

発行所　株式会社　三恵社
〒462-0056 愛知県名古屋市北区中丸町2-24-1
TEL 052（915）5211
FAX 052（915）5019
URL http://www.sankeisha.com

ＩＳBN978-4-86487-862-3 C8793 ¥1800E